SCIENCE IN POETRY

An Intermediate School Review

by

Winifred Schramm

SCIENCE IN POETRY

Cover & Art Work:
Taylor Springle
(281)992-0628

DEDICATION

*To my beloved family, Arthur, Marilyn, Noel, Allison, Craig
and the loving memory of Chris.*

ACKNOWLEDGMENTS

The encouragement of friends cannot be measured fairly by me:

Carolyn Perry, poet, critic, english teacher and friend, how can I thank you? Hundreds of lives have been enriched because you were their teacher.

Joan Kindred, science, nursing, and adviser, you have been an inspiration to me for many years, I value your opinions and your friendship.

Milana P. Suput, science, your studies and scholarship on two continents have served so many students. We are all better because of you.

All of you see poetry in science and science in poetry. I feel blessed.

"Work is the easiest activity man has invented to escape boredom." Anon.

A special tribute to all workers who have gone on before us.

INTEGRATING SUBJECTS

Over the years I have heard more than one scientist say, "I wish I had learned more metaphors and simile for my scientific writing." This inspired the writing of this book.

The world of poetry teems with figurative language.

I have worked with children in public schools for many years. I know that poetry can be taught using the subjects that children study.Now, language arts teachers can, if they wish, team with science teachers.

I hope this book will serve as a means to that end.

TABLE OF CONTENTS

UNIT 1: LIFE SCIENCES

Classifying Organisms

Plant Processes

Animal Kingdom Organisms

TABLE OF CONTE NTS (cont'd)

UNIT 2: PHYSICAL SCIENCE

Investigating Matter

Heat and Matter

TABLE OF CONTE NTS (cont'd)

SCIENCE IN POETRY

UNIT I: LIFE SCIENCE

Classifying Organisms

Cells

Cells make up all living things;
Fish and fowl, scales and wings.
Life processes demand food
Life cycles turn, and this is good.
On this wheel, the things we eat
Pump hearts, work brains and move our feet.
Cells break down to energy
For plants and animals and me.
Scientists know that ninety percent
Of the plants in the sea have food content
That is shed as waste to the ocean floor
For tiny animals to devour.
The cycles of energy are then released
Far from the sun; life forms are increased.
As new cells grow they replace the old;
Life cycles of all living things unfold.
When we study the plan we appreciate
How forms reproduce and regenerate.
Adventure in science will beckon, impel
When we open our world to the study of cells.
Anton van Leeuwenhoek
Used a simple microscope.
Since "scope" means see and "micro" small,
With this re-discovery we see all
Those cells that we could not see before,
A scientific tool that helps us more
Than most of us know. The things cells eat
Can add to our number while others delete.

SCIENCE IN POETRY

The one-celled plants and animals we view
Have openings for water and food to go through.
A scientist named Dr. Ernest Everett Just
Discovered these openings; all cells must
Be able to feed and pass waste out.
This process tells us what cells are about.
The causes and effects are within our scope
When we use the wonderful microscope.
The way scientists identify is simply to classify.
Three ways that are different . . . or alike to go by
The scientists list; when this is done
They place every organism into five kingdoms.
Cell structure separates plant cells from bacteria
The nutrition used for energy is another criteria.
Energy used by plants from a far-away sun
Would not feed all animals - and you and I are one.
The third way that scientists classify is use
Of knowledge . . . of the ways that organisms reproduce.
Organism is the name we give to living forms;
Tissue is made of like cells, specific tasks performed:
Define the work of tissues; consider organ brain,
Its nerve tissue sends messages out and back again.
Cells define an animal, cells define a plant.
Scientists may try to mix the two, but as of now, they can't!
Cells, tissue, organs and systems, systems we mention last,
They are a group of organs that do a certain task.

Life Process

Living things reproduce to make more of the same,
Dividing to make more of them is one life-process game.
Nutrition from the food they eat will help the cell life grow.
Growing is another must as all nutritionists know.

SCIENCE IN POETRY

Expelling waste is a must to keep live things alive.
Without this vital process, and this is number five,
The organism would die. Reaction is number six;
If we could not react to danger, we would be in a fix!

Five Kingdoms
Plants, Animals, Fungi, Monera and Protist
Monera has no membrane around its nucleus.
It takes in food . . . it is good and bad for us.
Examples: Monera turns milk into cheese
While others may cause disease.

All protist like wet places. Two kinds are mentioned here,
One-celled protist eat organisms which happen to be near.
Seaweed, a many-celled protist, gets nutrition from the sun.
A membrane is around each nucleus, whether many-celled or one.

Fungus is a kingdom that makes chemicals.
These chemicals digest what fungus grows upon.
Yeast, mold and mushrooms all are fungi, too.
I might add that penicillin was made from fungi brew.

Genus and Species
A genus is a group of animals that are alike in many ways.
A species means one of a kind; classification pays.
Scientists give scientific names in Latin or in Greek:
That way they know genus or species when they write or speak.
Dead languages are useful in the naming game.
You know if a language is dead the names will never change.

SCIENCE IN POETRY

Green Plants

In green plants life circles go around and around,
Sun and air from above and water from the ground.
Root hairs grow from the ends of each plant
They gather water and food, this plan is excellent.
There are two kinds of tubes in a plant's stem.
Food and water are carried up and down by them.
This process is called transpiration.
Through stomata, water is lost as in perspiration.
From the roots to the stem to the leaves through veins,
The cycle of processing food begins.
Stomata are openings on underside of leaves.
Plant food cannot be made without air these plants receive.

SCIENCE IN POETRY

Plant Processes

The process of plants making food is called photosynthesis.
Oh, what a wondrous gift to us, sunlight's constant kiss!
Of photosynthesis, we must say more,
As ink in a blotter, life seeps through each pore.
These use green-colored chlorphyll for food making tasks.
In plants, small green bodies are call chloroplasts.

Sunlight is trapped in these food-making cells,
That green wonder-substance we call chlorophyll.
A cell-wall protects and covers the cell.
A cell membrane has a function that it does quite well.
It lets food, water and gases exchange.
Plant cells are larger than other live things.
The nucleus controls the work in a cell,
Without this direction cells might grow pell-mell.
Tiny sacs of water help the cell hold its shape.
These are called vacuoles, they float in a cytoplasm lake.
So water comes into the leaf through veins,
Sun's energy breaks water into gases and then
The carbon-dioxide remains, so you see
The plant gives off oxygen to the world and me.
The air the plant uses has hydrogen,
This joins with carbon-dioxide and then
Sugar is made, our life-line of food.
Now isn't this plan far better than good?

SCIENCE IN POETRY

So sugar is made in the leaves, as we know.
Veins transport sugar so the plant can grow.
Then oxygen combines with sugar, you see,
To release a life-process called energy.
What is lovelier that a tree
That blows in the wind wild and free?

SCIENCE IN POETRY

Photosynthesis and Respiration

To know the difference between photosynthesis and respiration,
Let us study more rhymes for our contemplation.
Photosynthesis takes place in cells with chlorophyll.
Respiration takes place in all living cells;
Photosynthesis feeds plants with the sugar it makes.
Respiration uses sugar which it cannot make.
Photosynthesis stores energy as sugar from the sun.
Respiration releases energy sugar has spun.
Respiration produces carbon-dioxide, a waste.
If we didn't have oxygen we would vanish, post haste!
The miracle of plants must be hailed far and wide...
They give us oxygen for carbon-dioxide.

Reproduction

The word " reproduction" means
The continuation of all living things.
If plants and animals could not reproduce
We would all disappear without any excuse.
The flowers of a plant contain
Reproduction seeds of a like-kind grain.
The petals of a flower are often colorful.
The petals serve to protect the ovary and ovule.
The stamen of a flower is the male reproduction part.
The pollen-grain these attract must fertilize to start
The ovule growing in the ovary. Pollinations begin
When pollen grain goes downward into the pistil opening.

SCIENCE IN POETRY

Monocot and Dicot

If there are three petals in a flower
You have monocot seed in there.
Double three and you have six.
That monocot plays double tricks!
Four petals come from dicot seed,
Sometimes five petals fill dicot needs..
You just can't tell unless you know
The petal count the flowers grow.

Flowering Plants

The processes of a flowering plant may be viewed as these:
The pollen sticks on stamen when carried by wind and bees.
Pollen tube grows into ovule when stamen fertilizes,
This egg-cell becomes a seed. The fruit around it comprises
The meat of pears and apples and most of the fruit we eat.
Pollen and fertilization in plant life is very neat.

Conifers

Conifers make seeds in cones
Green trees and plants with leaves, sharp, needle-honed.
The cones are tough and they sheathe like a blade.
But the trees are lovely and they make good shade.
A pine tree is a good introduction
To a conifer's method of reproduction.
This is worked out to an efficient degree;
It grows its own pollen cones and seed cones, you see.
Examine the seeds in a cone of a conifer.
They are as light as a feather and they fly everywhere.
When the wind picks them up and lays them down,
Trees will grow and stay green all year around.

SCIENCE IN POETRY

Gymnosperm to the Greeks was a "naked seed" name,
It doesn't grow fruit or flowers but reproduces just the same.

Reproduction Without Seeds or Cones: Ferns and Mosses

A young fern plant makes eggs and sperm,
An interesting plan, as we will learn.
The rain water helps the sperm-cell move
To the egg-cell part; this plan proves
An efficient way to make these spores.
This begins new plants on our meadows and moors.
The ferns have tubes but mosses have none.
They feed from nutrients and water from the ground.
The difference in mosses and ferns, you know.
When fern-eggs are fertilized, new ferns will grow.
In moss, the fertilized eggs grow into spore balls
Which ride on the wind; moss grows when they fall.

Plants Without Seeds

There are plants without seeds or spores, we know.
The "eye" of a potato is the part that will grow.
Isn't it clever of the potato plant
To be able to do what other plants can't?
Many plants reproduce from root, leaf and stem;
Potatoes and onions are just two of them.
Water is needed for new "runner" stems:
These plants grow quite well with food under them.
Of course they, too, are fed by the sun.
Stem-grafting makes two when once there was one.
How interesting are plants, trees and their structure;
If you make them your work, you study horticulture.

SCIENCE IN POETRY

Animal and Plant Cells

We must compare cell difference:
To identify and make good sense.
The difference between plant and animal cells
Are choices of food and structure as well.
Both plant and animal cells have these:
Cell membrane, cell vacuoles, cytoplasm and nucleus.
However, a plant has a cell wall,
But an animal cell has no wall at all.
A plant cell has the chloroplast scheme
That traps the sun in its chlorophyll scheme.
There are four parts to animal cells and six parts to plants
These two parts in plant cells make the following difference:
Cell plants may be dried in a non-living wall,
But animal cells will rot as they have no wall at all.
We've mentioned the plant trapping energy from the sun...
If we had this process what work would not have to be done?

Animals: Two Main Groups

Vertebrates

The backbone is the vertebrae
That holds us up in a clever way.
This link of bones from head to seat
Make us quite rare and we think, neat.
We are rare because there are only four in a hundred
Of the animals in the animal kingdom
That have this important link called vertebrae.
Aren't you glad that we are made this way?

SCIENCE IN POETRY

Invertebrates
Sponges
The creatures without vertebrae
Have no backbone to swing and sway.
Some, like sponges soak and sit.
We call a sponge invertebrate.
They soak up water through their pores,
Thread-like whips bring water stores
Of food and oxygen; so energy
Is released in lake and sea.
Since they can't move, you might think
That they would soon be eaten, but they stink!

Hydra, Sea Anemone and Jellyfish
The hydra are invertebrates
That live in freshwater, streams and lakes.
Their hollow sac-like openings
Have tentacles around a ring.
The hydra, sea anemone and jellyfish
Shoot stinging cells at a tasty dish.
They paralyze small creatures for food
At exploding poison they are quite good!

Worms
A flatworm is called a planarian,
He can exchange his head for another one.
Do you believe that it would be great
For all animals to be able to regenerate?
A tapeworm is a flatworm parasite,
He steals digested food from a host's digested diet.

SCIENCE IN POETRY

A Round worm eats dead animals and plant matter.
He lives inside host: host gets thin, worm gets fatter.
Since other organisms form his diet,
We call this worm a parasite.

A Segmented Worm is round with rings, he has heart and blood.
He moves with thorn-like bristles through the soil and mud.
The earthworm is segmented, the one we know about
Hisbodyisrealshort until he s t r e t c h e s   o u t.

Echinoderms

The echinoderm has a hard, sharp spine:
On tube-like feet he does recline.
A sea-star is an echinoderm.
He grows body parts as a planarian:
If one foot is cut off, he can recover
And quickly grow himself another.
Just because he loses his head,
It would be folly to call him dead.
With a foot just like a suction cup
The echinoderm can suck life up.
I am inclined to think, time and again,
That I am glad most organisms are not planarians.

Invertebrates With Soft Bodies

The mollusks are invertebrates,
Some have soft bodies with calcium plates.
This covering we call their shell.
They have a blood-pumping system as well.
They have the beginnings of an eye.
I have never seen a scallop cry.

SCIENCE IN POETRY

Clams, oysters and scallops have two shells.
Inside these shells the mollusk dwells.
Each has a foot to move along -
This muscular foot is very strong.

Snails and slugs are mollusks too,
The foot gives off a slimy goo.
But slugs don't have a place to hide,
While snails with shells can hide inside
When enemies are seen about.
Poor slugs, they are nude and are food, no doubt.
The limpet has his shell on top.
A gastropod still clings to rock.
On a menu, if you see snails
They are called "escargot" and you will dine quite well.
Food prices rise with a foreign name -
They are snails, snails, snails, snails just the same!
Two mollusks show that they can think
Because they squirt large clouds of ink:
The octopus and the squid are great
In their ability to escape.
They both have feet like a suction cup
To catch their food and bring it up.
Each has a mouth that opens his head.
He does not care if the food is dead.
They grow quite large, but hold on . . . wait
I use those little squid for bait!

Those that grow to eighty feet
I hope I never chance to meet -

SCIENCE IN POETRY

Except in stories of the sea.
When I read I'm as happy as I can be!
Do monsters hide, do you think,
In writers' long, long spreads of ink?

Oyster nacre grows a pearl from one grain of sand...
Can irritation be a nature-trait, well planned?

Animals With Joined Feet
Arthropods

The arthropod is classified in the group, invertebrate.
His skeleton is grown outside, an exoskeletal plate.
Millions crawl, millions swim and millions fly.
Do you think that they could rule, or that they might try?
No, of course not, you must say.
The earth was never planned that way.
Number of legs and body parts help scientists classify.
Science must have set guidelines to go by.

Crustaceans

A Crustacean has five pairs of legs, his body parts are two,
Muscles attached to an exoskeleton move as our ligaments do.
Shrimp, crab and lobster are three we will name..
Male lobsters are quite vicious and killing is their game.
In food-stores you will notice the lobsters have a band,
With their two front claws they crush, they use them as hands.
They will kill everything in view:
Fish and other lobsters too.

SCIENCE IN POETRY

Arachnids

Ticks and mites are parasites. They have no use that I can see;
They suck the blood of animals, including you and me.
Four pairs of walking legs, their body parts are two . . .
Spiders are listed with them, although they surely do
More good than harm for people; However, an hour-glass
Upon a spider's back means you should kill her fast!

Insects

Insects outnumber us all as we might suspect,
With over seven hundred thousand species, we can't be derelict
In finding ways to stop them when they cause us harm.
Insecticides are big business for the ranch and farm.
Insects have three body parts: head, thorax and abdomen,
They also have three pairs of legs and wings.
How scientists know how insects see
Is certainly a mystery to me.
With compound eyes and so many lenses . . .
How they see flowers just staggers our senses.
Two feelers called antennae are on their heads,
They can taste, hear, smell and feel with these, it is said.
Insects are equipped with mouths that they need,
To cut, chew, suck or pierce when they feed:
When exoskeletons get too small they split their shells to go.
They molt and leave "exo" behind: they may be friend or foe.
As friends they eat the bugs that feed on every farmer's crop.
The trouble comes when they are foes and never seem to stop.
As foe they may appear in droves as grasshoppers will
To eat the crops and stubble too, before they get their fill.
A moth can eat a woolen suit or fill it full of holes,
His size denies this enterprise. We wonder where it goes?

Mosquito bites can raise a stone with a sad epitaph,
But most of them, thank goodness, only raise our wrath.
Is there some real good that from the housefly comes?
I cannot see, so get the swatter. Here comes another one!

Millipedes and Centipedes
Millipedes and centipedes have many body segments,
Millipedes eat plants, but centipedes' diet requirement
Depends on worms and insects, and they have claws and poison.
You like the millipede the best? If so, what is your reason?

Vertebrates Classified:
Fish, Amphibians, Birds, Reptiles and Mammals
The five main groups of vertebrates we catalogue:
Fish, amphibians, birds, reptiles and mammals such as dogs.
We've said that vertebrates have bones from head to seat
To hold them up. To move about, a backbone can't be beat.
Fish are cold-blooded animals: they match the temperature.
Scales protect, their fins propel, but they could not endure
If gills could not take oxygen. They have to breathe you know.
Exchanging carbon-dioxide waste, the fish can live and grow.
Isn't it great that this waste feeds the plants down there?
Those plants reciprocate. They give oxygen back to be fair.
Lateral lines protect the fish, tiny holes with nerves,
These, and a powerful sense of smell, alert, and often serves
A warning when enemies are around.
They are so lucky for they feel the pressure of a sound!

SCIENCE IN POETRY

Amphibians

The vertebrate amphibian
Lives in water and on land.
Toad, salamander and frog
Prefer land, especially a bog
To keep his tough skin slimy and wet.
I have not seen a bullfrog yet
Who did not have a jumpers's legs.
In water, amphibian lays her eggs.

Most amphibians have to breathe air
With lungs. But some, like mud-puppies fare
With gills so they can breathe beneath
The water surface. This is a feat.
Tadpoles are baby frogs, you know.
Now you should see those tadpoles grow!
As comma-like blobs they grow a tail,
With four legs and lungs they do quite well.
The pattern these amphibians keep
Makes old time rush and science leap.
If such is nature's awesome good
I would not change it if I could!
To tell you a story about frog's old age,
Please stay with me and turn the page.

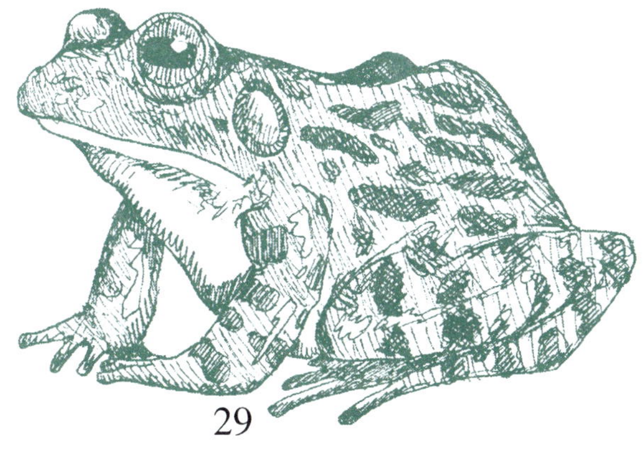

29

SCIENCE IN POETRY

Survival

The coelacanth is not extinct,
But none had seen this fish before
Except as fossil turned to stone.
Seventy million years or more
Had passed since he was on the scene.
Then, Africa's most southern shore
Gave him to scientists once more.
But coelacanth is very rare.
His family is crossopterygians.
His fins are paddles. He could move
But unlike rhipidistians,
He could not crawl out on land.
So "Coel" and "Rhip" each went his way,
Then "Rhip" developed legs, they say.

But "Rhip" it seems has disappeared.
And coelacanth? Well he lives on.
So oceans must be friendlier.
It follows coelacanth has won
If living on is species goal.
But frogs are "Rhip's" most distant sons.
We see him in amphibians.

1. coelacanth = se´-la-kanth
2. crossopterygian = kros-sop´-te-ridg´-e-an
3. rhipidistian = rip-i-dis´-te-an.
4. amphibian = am-fib´-e-an

Kind of Poem: This poem is a septet stanza which means a seven line stanza. This scheme is "abcbdbb."

SCIENCE IN POETRY

Birds

Birds are warm-blooded. This does not change
If temperatures rise or drop outside the normal range.
Like reptiles they lay eggs in shells to hold insides in.
Feathers are on birds alone so heat is kept within.
Feathers help the birds fly: these are strong and light.
Birds have beaks and wings they need for food and flight.
The poem below will try
To explain how birds fly.
(Couplets)

SCIENCE IN POETRY

Wings

Both birds and planes are enigmatic things
But planes, of course, are better understood.
Propellers and hot gases give planes wing;
Flight feathers propel birds as paddles would.
All birds have wings shaped to their evolved need;
Short rounded wings allow for sudden flight.
Without them, some small birds would, indeed,
Become the menu for some others' diet.

Long pointed wings serve distant voyagers,
Falcons, swallows, plovers, wild geese and swifts.
The largest wings belong to land-soarers;
Eagles, hawks and vultures ride the windy drifts.
The speed of flapping wings denote the size,
The larger wings move in slow rhythmic grace.
With our computers, pictures, formulas,
The gift of wings is still denied to us!

(Kind of Poem: This poem is called an octave. It is written in two stanzas of eight lines each in iambic pentameter. It has a heroic couplet ending. Heroic couplet means that the lines scan the same number of feet. Heroic couplet has a ten syllable count in each line.)

More Vertebrates
Lizards, Alligators and Crocodiles, Turtles, Mammals and Snakes.

Lizards have scales and most live on the land.
They look somewhat like dinosaurs, miniature planned.

They have thick scales and claws on their feet.
Chameleon lizards can repeat
The color or background that they are on.
An army should envy chameleons.

Alligators and Crocodiles
Alligators, and their cousins, the grinning crocodiles
Live in water and on land and they dine in style.
They live on fish or any other animals.
I saw one pull a dog down and I was appalled!
Alligators are black and gray,
Crocodiles are green and gray and may
Have more slender, vicious snouts.
Both know what eating is about.
They have four legs, they run and swim.
It is wise to be wary of them.

Turtles
A turtle is a reptile too, you know.
He carries his house where he may go.
He can pull his head and feet into his box.
This is called his shell, and he can outfox

Any hungry animal who wants to dine.
He pulls in his head and feet just in time.
Turtles are ancient, they can outlive us;
I think it's because they live without fuss.
Turtles lay eggs and leave them . . . do they live long
Because they are free from the moment they are born?

Snakes

Cold-blooded, dry-skinned, scale-covered snake . . .
My how you twist and undulate!
Snakes make up the largest group.
They don't have legs but they can loop,
Curve, and climb and swiftly crawl.
Some are large and some are small.
At 9 meters long, that is 27 feet-
An anaconda can't be beat.
The thread-snake is only 12 cms . . .
So you know that there are only 4.5 inches of him.
Some snakes can swallow the entire thing
Be it mouse, egg or bird . . . or in a ground-squirrel range.
Their jaws unhinge to open wide,
Their muscles squeeze when there is food inside.
I mean snake muscles move food down
They have to eat so please don't frown.
Their teeth are placed in a backward curve
Their prey can't escape, and so they serve.
What would you do if you were on this camping trip with me?

SCIENCE IN POETRY

Camping In Texas

I lie awake, the night is dark
The campfire hisses as it dies.
I saw him coiled there in the park
The fire's last glow had shown his eyes.
The campfire hisses as it dies,
If I cry out now will he strike?
The fire's last glow had shown his eyes
Twin terrors in the fire's last light.
If I cry out now will he strike?
No, no, I must lie quiet and still.
Twin terrors in the fire's last light
Hatred sent me thoughts of "Kill".
No, no, I must lie quiet and still.
I feel cold skin against my arm.
Hatred sent me thoughts of "Kill."
Perhaps, perhaps he'll do no harm.
I feel cold skin against my arm.
I saw him coiled there in the park.
Perhaps, perhaps, he'll do no harm.
I lie awake, the night is dark.

This poem is a pantoum. Notice the repeating lines.

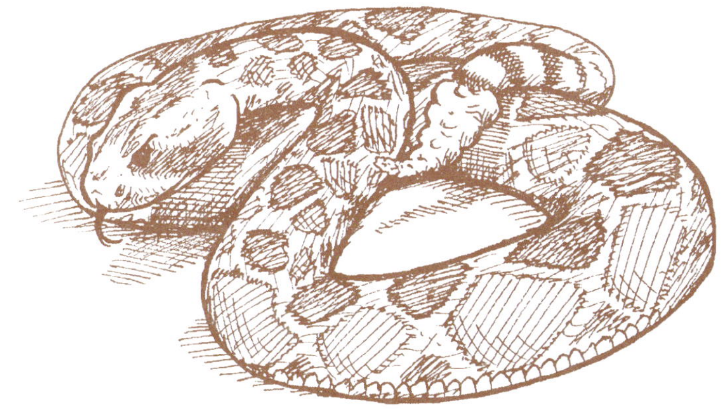

SCIENCE IN POETRY

Mammals
The fifth group of the vertebrates
Are mammals as our science book states.
They are warm-blooded and they make milk
To feed their young; some of their ilk
Have fur and hair and live on land.
A mothers produces milk from a mammary gland.
All have lungs and they must breathe air.
You will find some mammals everywhere.
The baby grows inside the womb-
Young kangaroos arrive too soon.
They climb into their mother's pouch
They kick from outside in . . . ouch!
These little things, they have to grow
Before they can move out, you know.
I'll bet that her relief would astound
When those baby's feet hit the ground!
Some mammals can be harmful ones. I'm thinking of the rat,
But we have ways to combat them, one way is mammal . . . cat!

Which group of vertebrates includes you?
I have to write about mammal dolphins before I am through.

SCIENCE IN POETRY

Our Friends, The Dolphins

Dolphins are such lovely creatures,
They even have some human features.
They must breathe air as we do,
By way of blow-holes they breathe through.
They hear the ocean pebbles move.
They swim in darkness fast and smooth.
Their heads are big, long, slick and flat
With pin-hole ears most keen for that.
Their sonar keeps them well aware
Of friends or enemies out there.
Our studies of them well might be
The answer for our blind to see.
The dolphins navigate the dark;
To them such freedom is a lark.
They whistle, squawk, bark and mew.
They like to be with humans, too.
They guard their young, protect their old.
They fight with cunning, I am told.
In intellect they rival us.
Sometimes they are ridiculous.
They grin and flip outrageously
When they perform on land or sea.
It has been said they rescue man.
They push him gently toward the land.
Of all the creatures of the sea,
The dolphins mean the most to me.

(Couplets)

SCIENCE IN POETRY

Populations and Communities

Scientists often estimate
The Number of things on a grid-plate.
They take the numbers in any two squares
And add for the sum of the pair.
Then they divide by two for an average,
The grid may be small or very large . . .
They multiply the number by the grids on the plate,
This way they have an educated "guesstimate."

Populations

A population is defined
As a species that are of the same kind
That live in one certain place.
A population could be of the human race . . .
A community, and we often list them,
As all populations in an ecosystem.
An ecosystem may be of non-living things
Such as soil, water and buildings.
How they affect the living, you see,
Is called by the word ecology.

Environment

An ecosystem tries to meet every need
That an organism has, whether plant or seed . . .
Or an animal that another organism eats.
A habitat is where the organism retreats.
A habitat is the organism's home,
He might end up as prey if he did not have one.
A field mouse has a tunnel for his habitat.
The bird has a nest, though there is danger in that.

SCIENCE IN POETRY

A habitat helps the animal stay alive.
A nitch is the purpose or the way species thrive.
Your home is your habitat. Your nitch, right now, is school.
If you study and work you will do well, as a rule.
A community must have a plan
To feed the population and
Determine how populations grow.
We learn from natural plans, you know.
Consumers, producers, scavengers, and decomposers
Each have a role in the bills of fare.
"Bills of fare" means the menu
Or "what organisms eat" would explain this too.

Water plants and algae are producers in a lake-
Tadpoles and snails are consumers, for they take
Food from the plants . . . other consumers eat them
So everything has a place in a food chain.

A crayfish is a scavenger. He eats dead animals and plants.
Bacteria and fungi decomposes, there is also good in that.
There always was competition for food: No one can deny it,
Sometimes it comes down to "survival of the fittest."

If the birth rate is too large, many organisms die.
Nature tries to balance birth and death with food supply.
When food, water and space are scarce, populations decrease;
Overpopulated countries have famine and disease.
We need to try to help them all with food and medication:
Our work will never cease, for they too, need education.

SCIENCE IN POETRY

Endangered species are the ones where few organisms remain.
Extinct means that the organism can't come back again.
Extinct animals become fossils, bones are all that is left.
We must protect our animals. Good environment is our wealth.
Successions are series of changes that will take place
When nature repairs damages when fires and flood efface
The trees, plants, and animals that once were living there.
Nature is always busy with destruction and repair.
When winds blow seeds and rains come down
Grasses, plants and trees rebound.
This first stage, the pioneer stage,
Could take years or even an age.
Insects reappear, the first eat the grass,
Then birds eat insects; all are compost at last.
As bushes and young trees grow in this land
Larger animals appear and once more make a stand.
When at last a new forest is in evidence,
It rose from the ashes like the fabled Phoenix.
This forest will be a climax community.
Does this fact say anything for continuity?
Succession slowly changes a community population.
Food chains and environment have a close relation.
To animals that live there, a filled-in pond
Becomes a rich environment; new animals respond
To grass, to trees, to farmlands decayed matter makes
Good soil for food to grow in. Nature gives and takes.

SCIENCE IN POETRY

UNIT 2: PHYSICAL SCIENCE

Physical Properties

Mass is the amount of material in an object.
Volume is the amount of space the object takes up, in fact
All matter has certain properties...
As color, smell, state and taste are properties of cheese.
Were you to melt or slice the cheese
This would not change cheese properties.

Chemical Changes in Properties

When one kind of matter reacts to another kind,
A chemical property is involved and you will find
That chemical changes come about as in burning wood.
The gases and ashes from burned wood may be bad or good.

A fireplace is cozy, a burning house is bad . . .
But chemistry is always working to make us glad or sad.
Four kinds of matter are now known,
Solids, liquids, gases and plasma, the newest one.

Atoms

A model of an atom explains what it is like;
It is much too small to see: you need a powerful "mike".
The microscope shows bits of matter; let us mention some:
An electron has a negative charge, a neutron has none.
The proton has a positive charge, 1800 times larger than
The electron. Protons have a gigantic span.
The protons and the neutrons form the nucleus.
This nucleus is small indeed, compared to atom's mass.

ERIODIC TABLE OF
THE ELEMENTS
POETRY
POETRY
PRINGLE

SCIENCE IN POETRY

Most of the atom is empty space around the nucleus.
Electrons travel like space ships around this empty mass.

Elements
Scientists know that there are 110 different elements.
Now some are made in laboratories, their contents
Are neutrons, protons and electrons, of course numbers vary.
Elements are identified by the kind of atoms they carry.
The simplest element we know has one electron and one neutron.
The most complex element has 92 electrons and 92 neutrons.
An element is a substance that cannot be divided.
It is matter in its simplest form, this fact has been decided.
Discoveries continue. Could old Time change this?
A Periodic Table carries the latest list.

The atoms in an element give matter its name.
Gold, copper, silver, iron and tin are a few that we can name.
Most elements are metal and they conduct heat;
Conductors carry energy and give us light and heat.
Most non-metals are solids and gases at room temperature.
Sulfur and carbon are solid non-metals we must mention here.

Oxygen and nitrogen gases are in our compound air . . .
Without these non-metal elements, how would we fare?
Silicon is an element with properties we know
That are good to use in computers, radios and videos.
Think of it like fine white sand, debris of rocks and such.
We know that it is plentiful, one fourth of the earth's crust.

SCIENCE IN POETRY

When two or more atoms join they form a molecule.
That is how we learn to recognize matter as the rule.
A compound is a substance when different atoms join.
Water is a compound: hydrogen and oxygen.
We have studied about carbon dioxide throughout this book.
Since CO_2 is the compound, you know with one close look
That there are two parts oxygen and one part carbon here.
Every molecule of this compound is the same in kind and number.
These symbols are our short-cuts to name the elements.
The formula shows which elements are in matter contents.

Grouping Elements

Thus scientists have been able
To make a periodic table.
Elements may have similar properties,
The number of protons determine these.
Symbols are the keys that they use,
So elements are not confused.
We have said that when two or more atoms join
You will find a molecule is formed.

Solutions and Suspensions

A solute is a substance that dissolves
When mixed in a solvent. This fact proves
You can make a solution of Kool-aid when you
Mix Kool-aid powder with water. That is all you have to do!
A solute may be solid, liquid or gas.
The gases in drinks may escape very fast.
You can prove to yourself the truth in that
When you taste a drink and declare, "This soda is flat."
Solutions may be either weak or strong:
Amounts of the solute determine if they are right or wrong.

SCIENCE IN POETRY

Dilutions indicate the strength of the mix.
Concentrated means more of dilute in the mix.
Suspensions are liquids that soon separate.
An oil and vinegar dressing will instruct " shake".
The particles of mixtures may be very small
And stay so scattered you won't see them at all.
Thus a solute dissolves in solution,
But a liquid that doesn't is a suspension.
For you to know and be made aware
Remember, "in suspension, particles just hang in there."

Heat and Matter

The ratio of Centigrade to Fahrenheit
Will always help you read a thermometer right.
With (C) Centigrade water freezes at zero..(O) degrees.
In (F) Fahrenheit water freezes at thirty-two (32 F) degrees
In Centigrade water boils at one-hundred (100 C) degrees.
Fahrenheit two-hundred and twelve (212 F) is boiling, you see.
To change C to F is like climbing a tree.
37 (C) is body temperature. Multiply C by 9/5 and add 32
C x 9/5 = 333/5 = 66 3/5 + 32 = 98.6 F.. . you are through!
To change F to C , (going down the tree)
98.6 F - 32 = 66.6 x 5/9 = 333/9 = 37 C
Body temperature is important to diagnose disease.
Thermometers of different make are used to know degrees.

Mercury and alcohol expand inside a glass tube.
Electronic and liquid crystals are also now in use.
Pyrometers can measure heat too high for these mentioned here.
They measure heat by colors of light, and bring science near:
To outer space, to ocean depths, to age and heat of a star.
Scientists can measure distance and stay where they are!

SCIENCE IN POETRY

Matter is made up of particles and molecules,
Thermal energy is how fast or slow they move.
Heat is the flow of energy from cool to warm.
Temperature and heat determines good and harm.
Metals are matter that is hard to beat
When it comes to the conduction of heat.
Gases are not good for heat conduction,
Else the world would be in a sad condition.
A simple deduction would soon prove this:
If gases were conductors, we'd burn to a crisp!

Materials that do not conduct heat or electricity
Are called insulators. Let me name a few of these:
Wood handles for pots and pans, your thermal underwear,
Your outer-cover- "skin," even your head of hair!
Trick question: Can you guess how you insulate
Yourself from a big science test mistake?

Convection
Do you think that it is a surprise
That cool air sinks and warm does rise?
Convection is the word they say
When a room is warmed in this heat cycle way.
Convection proves the work of blends
Outside this circling make our winds . . .
Energy that travels through an empty space
Is called radiation, which surely takes place.

Expanding Matter
In the rays from the sun: in our microwaves,
Just think of the manual labor radiation saves!

SCIENCE IN POETRY

When you heat matter, that matter expands.
The proportion it expands always depends
On whether it is solid, liquid or gas.
The solids, of course, expand the least.

Liquids expand more than solids, you see;
Gases expand the most of the three.
The opposite happens when matter loses heat.
When matter cools, those little molecules retreat.

Contraction

When matter contracts it takes up less space.
There are exceptions of course, as in the case
Of heating rubber or freezing water to ice:
Rubber contracts and the water expands to make ice.

But usually heat causes matter to expand.
While cooling causes shrinkage, like the pores in your hand.
If you think expanding and contractions are weak, think twice:
Water freezes into glaciers and rocks break from ice.
Since there are exceptions to many rules
We need to study those crafty molecules.

Types of Energy

Don't waste potential energy in idleness!
Thus by now you must have guessed
That energy is what allows us to work
Just as electric energy lets the coffee perk.
Potential energy in matter is energy stored up.
Kinetic energy is an act, like filling up a cup.

SCIENCE IN POETRY

Mechanical energy is energy stored en masse . . .
The higher the distance from the cup the bigger the splash.
You have learned that particles in matter move.
This Thermal energy, or heat, goes to prove
Kinetic energy of all others is thermal by name:
Know "Thermo" means heat and you're ahead of the game.
Remember those little electrons circling in a cell?
They can march in a straight line as well.
This fact has proved most fortunate for you and me
Because these little soldiers are called electricity.
Electric energy, at present, could be called "the King."
But futuristically we may have to re-think our "energy" thing.

The sun is our answer, since heat radiates here.
Through oceans of space radiant energy appears.
Radiant energy is used in radios, sun panels and x-rays.
We must harness more and transmit; we must study ways
To be more efficient. Dwindling fossil fuels and population
Can further endanger our world situation.
We need more and more scientists: Could you be one?

Chemical energy is released by chemical changes;
Cars move and gas burns; This fact rearranges
The liquid of gas to gas in the air
So we have pollution and we must take care.
Car-pooling and less waste are things we can do.
Remember, the environment is made by me and you.

Nuclear energy is made when the atom joins or splits.
Our nuclear bomb was made as a result of this.
Thus nuclear energy is gleaned from the nucleus,
This can be good or equally bad for us.

SCIENCE IN POETRY

Energy Changes Forms

Any form of energy can change to a different form.
All plants change sunlight, radiant energy, from the sun
To chemical energy, sugar; so when we eat plants
This energy becomes mechanical when we walk or dance.
Energy is always doing something in the body, you know.
Even when you are sitting still heart pumps and blood flows.

Producing Electric Energy

Electric energy is carried by electricity.
The motion of neutrons, protons and electrons in unity
Becomes a current charge with heat.
You have seen electrons in lightening make a jagged streak.
Electrons are the negative charge in an atom.
Protons are a positive charge and they fight for dominion.
The neutron has no charge; it is a part of the nucleus.
The study of these three charges is important to us.

The pushing and pulling of electrons and protons
Is called electric force and this goes on and on.
Electrons move in many different ways:
Sparks may fly to you from cats and rugs when electrons play.
Electric current is somewhat tamed when it is in a groove:
Covered by insulators, it runs smooth.
Metals are especially good conductors for electrical flow.
But glass, rubber and plastics, you must know,
Insulate to keep electricity in, but some electricity is lost,
This raises electric bills to a higher cost.

SCIENCE IN POETRY

Dry Cell Electric Energy

A flashlight uses dry cell batteries
Which change chemical energy to electric energy.
A dry cell has a metal case
That holds a carbon rod and acid paste.
The chemical reaction between paste and metal
Causes electrons to flow and energy electrical
To come out as light.
Aren't we grateful for this on a dark night?

Wet Cell

A wet cell battery runs your car.
There is liquid acid poured in there.
There are dangers in a car battery
So be as careful as you can be.
A generator in a power plant produces electricity.
And turns mechanical energy into electrical energy.

A closed circuit is the path through which currents flow.
Electric circuit has three parts our banks of knowledge show:
First, a battery or generator to push electrons,
Second, a wire that the electron current rides on,
Third, the destination, or the final use
Whether it is to light a bulb or roast a Christmas goose.
The circuit will be closed when you turn the switch to "on".
Out comes the light or heat from helpful electrons.

We know that matter is energy.
It can masquerade as a genie, you see.
Electric energy becomes changed with heat internal,
In a toaster to become an energy called thermal.

SCIENCE IN POETRY

When toaster wires get red-hot
It becomes radiant energy on the spot.
Your electric stove acts the same way. . .
You can trace those electrons at their play.
The "light-bulb" story is what science is about.
We wonder how many filaments Edison burned out.
If the wire isn't thin to slow the electrons
The bulb will break or the filaments burn.
From electrical, to thermal to radiant
He had to have years to experiment.
When you say the word "mechanical". . .
You see a machine, motor and all.
The motor is run by electrical energy
That does the work mechanically.
That mechanical energy can shave a face,
Move loads of brick from place to place,
Mix foods, vacuum floors and dry your hair.
Three cheers for electrons and mechanical power!
Power is energy divided by time.
A watt is a unit for measuring power. This rhyme
Invites you to look at the labels on the tops
Of light-bulbs for the number of watts.
A twenty-five watt bulb makes a light too dim for you to read:
For the care of your eyes, a hundred watt is what you need.
A kilowatt-hour is a thousand watts used in an hour.
You can read your watt-meter to know your usage of power.

Each month the light company sends you a bill,
If you don't pay it, the company will
Cut off the power. So, turn off unneeded light!
Conserve that energy! Treat the company right.

SCIENCE IN POETRY

Electric energy is friend or foe.
You can be killed by what you don't know.
Water conducts electric heat:
Don't touch it with wet hands or feet!
Worn or frayed electric cords can cause a fire:
Respect electricity and handle with care.
Remember, knowledge, too, is power.
Too many "plug-ins" can cause a fire.
Fuses and circuit breakers help a lot.
If trouble comes, call an adult, an electrician you are not!

Work and Force

For work to be done an object must move.
The needed force is called push and pull.
A newton is a unit for measuring force.
This unit was named for a scientist, of course.
Force times distance measures your work.
A joule measures energy you use when you work.

Energy Resources

Think of all the plants and animals of a million years ago,
Plants packed down with sand and soil to form our seams of coal.
The big and little animals became our fossil fuel.
Crude oil from the ground is from this fossil pool.
So often when we hear a grinding and an engine roar,
Are we really hearing the last call of a dinosaur?
Crude oil is changed for our use in a refinery.
Our gasoline, jet and diesel fuel are from crude oil, you see.
Those fossil fuels from long ago make a contribution
We know that scientists must go and look for other solutions.

SCIENCE IN POETRY

Of course our coal and natural gas do not need to be treated:
Our modern world is much concerned that these will be depleted.
Power plants use fossil fuel; most of them use coal:
It's cheaper and the heat from coal makes the water boil.
The steam from boiling water makes the turbines turn.
The magnet in the generator does a constant churn.
The wire coil that wraps the magnet lets the electrons flow.
Electricity is the final product of this entire show.
Since carbon is an element, coal is from vegetation;
We know this heat releases gas that gives us hesitation.

Carbon dioxide is the gas that fires from coal release:
Carbon dioxide and air moisture traps a lot of heat.
This combination world wide has caused our world to warm.
If lands become too dry for crops, we face a lot of harm.
With growing population and greater need for food. . .
Our attitudes will have to change for our world and our good.
You all have heard, I do suspect,
About the heat-trapped "green-house" effect.
Gases released from fossil fuels result in acid rain
Which has the potential to kill every living thing!
Electricity is the king of most of our power forces.
Coal and oil, the fossil fuels, are non-renewable resources.
Remember, our ecology is up to you and me. . .
We want our world, city, and home to be pollution free.

Atoms As A Source of Energy

When the atom is split, energy is released.
Fission is the name of the process, dangerous waste increased.
When atom's nucleus is split, we call it nuclear power.
This splitting takes place in a reactor.

The heat this nuclear power makes turns water into steam,
As in coal and oil processes we are after electricity again.

Uranium is an element compared to coal and oil,
Which in the least amount will make the water boil.
Coal is used most, oil is next, then uranium. . .
This, too, can be depleted, and so the search goes on.
The waste produced by nuclear plants is dangerous.
Scientists are working constantly to make it safe for us.

Moving Water As A Source Of Energy
From steam to turbines to generators forces go.
Coal, oil, nuclear plants make electrons which we know.
Hydroelectric plants are powered by water falling down.
You find these plants near rivers, like Niagara, or a dam.
The falling water turns turbine blades just like a giant fan,
The generator's carbon coil sends electrons, which can
Produce enough power for New York, cheaper electricity.
Meanwhile a dam stores water, a good utility.

Tidal Energy
The energy we get from tides, whether high or low,
Is also made by locks and gates that lets in ocean flow.
The gates are closed and water is then higher.
When tide is low, the gates open, and falling water makes power.
As in the other plans we know, this plan is made to be
Another way our modern world makes electricity.
We know that other system fuel are nonrenewable,
But moving water is a force that goes on and on.
With dams the water changes the land:
In time land fills up with mud and sand.

SCIENCE IN POETRY

High tides happen only twice a day:
We can't make enough electricity that way.
Dams that block rivers can do much harm:
Fish, such as salmon, need their rivers to spawn.
Nothing is perfect, as we can see
In our relentless search for electricity.
Would it be good if we could store it up
In a non-conducting world-sized cup?
How would you decide what needed it most,
If it were rationed at reasonable cost?

Use of Sunlight
Solar energy is energy that radiates from the sun.
If a window faces the sun you will find it warm.
Because the sun causes the wind,
We must call sun and wind our energy resource friend.
The solar collectors we see on a building roof
Are glass-covered metal that offers us proof
That plates heat the water that flows through to tanks.
So we do use the sun when we don't have cloud banks.

But then in the places where clouds hide the sun
There is not enough heat to make the sun's energy number one.
Do you think it will be possible for mirrors to reflect
All the energy we will need from a spacecraft deck?
Mirrors are used now to deflect the sun to a spot
That boils water into steam, so hot
That metals can be melted; the resulting steam
Again turns turbines. This electricity is clean.

SCIENCE IN POETRY

Solar cells are used in our spacecrafts out there.
We cheer, we hope, and we want solar-cells here.
It would be my most fervent wish
That we could have electricity from a solar dish.

Solar cells change solar energy without turbines
Into electricity for rurals and urbans.
Our spacecrafts run on these solar cells,
This has opened up the universe: our knowledge banks swell.

SCIENCE IN POETRY

Wind

For thousands of years people have used the wind
To bring water to the surface and to grind grain.
A windmill is really a turbine, and you
Have studied the work that a turbine will do.

Geothermal Energy

In some of earth's surface, water turns to steam:
Geothermal energy is wonderfully clean.
This steam is trapped to run turbines in a power plant.
The trouble is these "hot spot" places are far too scant.
The distance electricity would have to go would be
An important cost-factor for this kind of electricity.

SCIENCE IN POETRY

UNIT III: EARTH SCIENCE

Earth's Changing Crust

The earth has layers that you can compare
With a peach. The seed would be like earth's core. There
The mantle of the earth would compare with the part you eat.
The crust of the earth compares to the peel of the peach.
The core is quite hot, made of solid iron.
A melted mass covers it, a molten mass of iron.

The mantle is the thickest, made mostly of rock,
Again, a molten mass of melted rock on top.
The crust is the thinnest, like the skin of a peach -
Solid rock that changes like sand on a beach.

Physical Weathering

Water, ice, and wind still shape
The mountains, valleys and shores: none can escape
When natural processes make their move.
They freeze, break, make rough or smooth.
These architects will always bring
New shapes from earth's constant weathering.
Physical weathering will not change
A rock; its chemistry remains the same.
When water freezes in a rock, the rock breaks down.
Soil remains; call it ground. Is this adjective, verb, or noun?

Do you think that a fragile plant
Can lift a weight that a human can't?
Consider a sidewalk near a tree:
Roots have pushed up where the walk used to be.

SCIENCE IN POETRY

Concrete gave way to a stronger force,
But physical changes take time, of course.
When wind and the cliffs play games all day,
The water leaps up and the rocks give away.
Wind and waves roll rocks and smooths them with sand.
This "sand-paper" action makes wind and waves into a hand.

Living things also bring about physical weathering.
Plants, animals, and worms break up soil by tunneling.
Man changes the land with roads, villages, and farms.
Coal and copper mines cave in and cause much harm.

Chemical Weathering

Physical weathering changes only matter's shapes and sizes.
Chemical weathering changes minerals of which matter comprises.
After you study the effects of the two
Which weathering would more quickly affect you?

Chemical weathering changes minerals as from acid rain,
The water, acid, and oxygen cause rocks to change.
An old car rusts when left outside,
Since minerals like iron plus oxygen make iron-oxide.

The same is true with ore in rocks or iron in the ground. . .
The soil changes color to yellow or red-brown.
Rain and water combine with carbon dioxide to make acid.
Which soaks into limestone rock to form caves amid
Places of limestone like Carlsbad Caverns whose means
Come from limestone and underground streams.

Calcium carbonate builds up to form stalagmites on the floor.
From the top stalactites drip to form more.
Moss and lichen grow on rocks to cause chemical changes.
The acids again break up rocks and nature re-arranges.
Mosses and other tiny plants
Cause chemical changes by root implants.
The chemicals that these small roots bequeath
Dissolve parts of rock as acids do in teeth.
Physical and chemical weathering are in place everywhere.
Wind, ice, and water move matter from shore to shore.
Which one is the stronger, do you think, and why?
Carefully list your reasons; you have a world to go by.

Erosion

Erosion is the name of change
When wind, ice, or water re-arrange
To move matter from place to place.
Erosion changes the earth's stony face.
Moved along by gravity
Water runs from the mountains to the sea.
The rain breaks up large clumps of soil;
Melting snow adds to the roll
That water brings to rivers and lakes.
Along the way the water takes
Tons of soil and surface contents.
Erosion changes continents.

Farmers plant crops in a certain way
To help hold the water in place and may
Prevent erosion of the soil. Terracing will
Help prevent soil erosion on mountain and hill.

SCIENCE IN POETRY

The cutting down of a river bed
Causes canyons to form. Overhead
The sides of cliffs are high and steep.
Grand Canyon is over-one mile deep.

Ocean waves erode the shore,
What once was land is ocean once more.
When rivers flow they deposit sediments,
So deposition means dropping contents
Like rocks, soil and moving sand.
My, but we have a restless land!

Sediments often enrich the land;
A farm by a riverside can be grand.
A delta is formed in the mouth of a river:
The depositions build from a water giver.
The river will give and take away.
One really doesn't know from day to day
If sand-castles that you build on the beach
Came from China, England or a farther reach.

Runoff water drains the land.
It branches like fingers from a broader hand.
These little run-offs become streams
Which roll and sweep the landscape clean.
Huge waves can crash and change a shore:
What once was beach is there no more.
Tides are force "that wait for no man -"
The story of time is the story of land.

SCIENCE IN POETRY

Glaciers and Wind Change the Land

The Ice-Age lasted ten thousand years.
The glaciers, ice masses, often appear
In oceans that border our earth's two poles.
Glaciers have played important roles.
The weight of the glaciers makes them move.
We see mountains and valleys where they once grooved.
There are two kinds of glaciers: one is an ice-cap.
Two-thousand square miles are in South Pole's white wrap.
The second kind forms in mountain valleys of snow
And looks like a great white river that does not flow.
We owe glaciers a lot: Our Great Lakes were caused
From glaciers that stopped and subsequently thawed.
Glacier National Park in Montana, USA.,
Has glaciers, mountains, and canyons; beauty all the way.

Wind

Wind can be a desert dancer, water dancer, or a fiend.
It can blow hot or cold as an enemy, or cool as a friend.
It seems to take sand in its handless hand
To scatter the sand like the devil's own fan.
In spite of the misery wild winds have blown,
I like its sure strength when I'm sailing alone.
The sand dunes it makes are fun to play in,
The kites that I make all run in the wind.
Some deserts are paved from the pebbles wind drops.
Desert sand will stay put when pavement holds rocks.
Wind will move farmland, beaches and such.
Do you trust the wind? I do, but not much.
Sandstorms can blind and windbreaks can save -
We must harness the wind, for it works like a slave!

SCIENCE IN POETRY

How Rocks Change
The crust of the earth has about twenty plates,
Which slowly move over the melted mantle-top, science states.
These plates may slide past each other, or they may bump;
When you see a mountain, you see the plates' hump.

When plates pass each other, melted rock may spew through,
You all are aware of the damage volcanos can do.
Earthquakes happen when crusts slide around:
A heaving fast tremor can flatten a town.

Most changes come slowly; some take a million years;
Then suddenly, scant warning, a mountain appears!
Rocks, too, are changing; rock cycles begin
When magma, which is melted rock, comes up again.

Igneous rocks are from magma's hot flow;
We have studied how rocks are minerals so,
If magma cools slowly, large crystals are made.
When magma reaches surface lava rivers invade
The mountains, the valleys, the cities and towns;
This is how the ancient city, Pompeii, went down.
Granite and basalt are two kinds of rock;
Granite is grainy, basalt comes from a volcanic drop.

Sedimentary Rocks
Metamorphic rocks are the results of change.
When plates of earth move, the rocks rearrange.
The pressure changes granite into gneiss:
The change in rocks' minerals causes this.

SCIENCE IN POETRY

Heat and pressure cause changes of many kinds.
Limestone becomes marble over eons of time.
Shale becomes slate; we write on slate surfaces.
We use slate for roofs, floors, and other good purposes.

The word "metamorphose" always means change:
Whether frogs, butterflies, or rocks, it is nature's domain.
Living things usually rot when they die,
But sediment cover can petrify.

Minerals replace dead molecules.
Fossil remains give us many clues
Of living organisms of an early time.
Teeth, bones, and shells are what we usually find.

When some trees died they turned into stone.
Our "Petrified Forest" is an example well known.
Sediments covered these trees long ago,
Water carried minerals, the changes were slow.

Those minerals replaced the trees' plant cells,
The trees resurfaced when earth's plates swelled.
A shell leaves a mold that looks like the shell:
When minerals filled in, casts formed where they fell.

An imprint is a track made by an organism;
A dinosaur track is just one of them.
The earth is surely one huge laboratory
And science keeps unfolding its wonderful story.

SCIENCE IN POETRY

The shells we have found on mountain tops
Remind us that earth plates cause flip-flops.
In silent voice the fossil speaks
Of oceans that once were mountain peaks.
It tells us of life so long ago!
The story continues: On with the show!

The Environment

We all have a need for clean air, water, and land.
We must teach all people of earth to understand
That sewage causes bacteria that can make us ill.
Pollutants can also damage or kill.

Some insecticides are harmful chemicals,
They can wash through the gardens and through farms fields.
They are eaten by protists that leave residue,
Mercury, harmful chemicals, food-chains broken through
Can end in our kitchen, so we have to find out
What chemicals we are eating. Science study is our scout.

Mercury in our system does harm to our nerves,
Ignorance is not "bliss"; knowledge preserves.
Thermal pollution will often impair
When heat is added to water or air.
Much harm is done to many living things
When heat is added beyond normal range.

Marine biologists take samples of water everywhere
Testing temperature and content of water we share.
Manufacturing must be careful as you can see,
When making products that can change chemistry.

SCIENCE IN POETRY

Oil spills are a menace of far reaching scope.
We will learn to contain them in the future, we hope.
Sixty times more water is stored in the ground
Than all other water on earth that is found.
This groundwater is the water we drink:
When rain and snow falls, these waters sink.
It goes to rock level, above porous rocks
Which keep water for us in quantity stocks.
We all must work hard to keep out pollution:
Nations working together is our only solution.
Farmers have stopped using harmful chemicals;
Sewage treatment plants cleanse water as well.
Insects eat others that can damage a crop.
A search for better farm ways goes on non-stop.

Breathing

You take twelve breaths in a minute, sixty minutes in an hour.
Seven-hundred twenty breaths an hour is your breathing power.
Seven hundred and twenty breaths times twenty-four in each day
Over seventeen-thousand breaths; clean air is a must, we say.

Sources of Air Pollution

Fires started by lightening can do much good,
Especially in forests, where they burn out dead wood.
Lightening fires happen, or shrubs and small trees would choke;
And of course these cause air pollution from both ash and smoke
These natural pollutants, volcanic ash and dust
Have happened over centuries. There are stories in earth's crust.
Natural pollutants may be dust, gas, or ash
From wind, fire, and volcanoes; but unnatural causes surpass

SCIENCE IN POETRY

In the harm that we suffer from sulphur and carbon monoxide.
Coal, oil and natural gas are used far and wide.

Unnatural pollutants, cars, especially
Burn gas and this pollutant harms constantly.
Carbon monoxide escapes to the air:
Our lungs must work harder when this gas is there.

All of these pollutants cause allergies or illness.
And wind-blown pollen can bring much stress.
Sulfur dioxide can join with rain
To make acid rain, which kills forests, crops, and everything.

Sharing rides to work cuts down on air pollution.
If people work together we can find a solution.
Auto makers make engines that use lead-free gas.
Lead pollution will thus be less and less.
Scientist constantly work to take sulfur from coal.
Plants use scrubbers for this worthy goal.

Land Pollution

Waste is a constant problem, since some won't break down;
Plastic, some paper, glass, and metal are a few well known.
Bacteria and water make good use of food.
Paper waste that can break down is also very good.
Open areas for trash are bad to look upon.
Getting rid of trash properly is a must for everyone.

SCIENCE IN POETRY

Toxic Waste

Say this, "Toxic waste is an enemy
To plants and animals and me."
We need to think of the growing things
That need clean air and fresh, clean rain..
Since our use of energy causes pollution,
Many steps can be taken to provide a solution.

Knowledge of conservation should be number one.
So begin to recycle aluminum.
Buy products in recycled paper used time and again.
Help the environment. This must be our plan:
To buy products that are contained
In recycled packages; Order is maintained
With plastic coated barrels and plastic enclosures.
These are ways chemical plants try to keep down exposures.

For every tree some companies cut down
A new one is planted in the same spot in the ground.
Reclamation is the name of this process.
This is a good plan and the damage is less
Than if they neglected to plant new trees.
If they fail, land erosion and floods, by degrees
Carry off rich soil and ruin cities below.
Prevention and conservation are good words to know.

Constant loud noise can damage an ear.
Noise pollution is worst where airports are near.
Laws must be passed to zone the land.
Airport locations must be well planned.

SCIENCE IN POETRY

Laws are made by us, and for us, you see.
That is the meaning of democracy.

Can you plan a town for folks to live in?
Draw stores, parks, schools and houses and then
You must include offices, hospitals and roads.
Think of highways you would build and the weight of loads.

Would you let eighteen wheelers drive on every street?
Would you pass laws on water to keep it fresh and sweet?
Would you let people throw trash anywhere? . . .
You might step on toes. Some have "freedom from care."
No, I will not tell you what you must write down.
But we will discuss it, what you want in your town.

Climate
The point of a curve that is closest to light
Will receive the most heat and be the most bright.
Climate means the average in weather conditions.
As affected by the sun to earth's changing positions.
Weather may change from day to day,
But climate remains in a most constant way?
Two features make up the climate of a place,
Temperature and precipitation are climate's true base.

Temperature is measured heat;
Precipitation is rain, snow, hail and sleet.
A hot-dry climate will have little rain.
It may be sandy desert or an arid plain.

SCIENCE IN POETRY

In a wet, warm climate it may rain each day.
The humid tropics are described this way.
Every climate begins with solar energy.
The sun is the source of greatest power, you see.
A layer of gases surrounds our earth;
This layer is called atmosphere; it is worth
Our time to read and understand
The purpose of this atmospheric band.
The atmosphere absorbs or reflects heat into space.
The atmosphere absorbs 20% of the heat in its embrace.
The atmosphere reflects 30% of heat back toward the sun.
The earth absorbs 40% of heat and reflects 10% in its orbit run.
In surface reflection, the heat waves warm the air,
But in that reflection carbon dioxide is caught there.
This gas has caused a condition called "the greenhouse" effect.
This trapped heat can do much harm and this we can't neglect.

If the earth became too hot, would the poles melt, do you think?
Would oceans rise and all of earth's surface sink?
The climate zones are interesting and we are fortunate
To be in an upper-middle zone that is named temperate.
The polar climates are quite cold; sunlight spreads far out.
Proximity to the sun is what climates are about.
The equator is of equal distance from the North and South poles.
The tropical climates are quite hot because of equator's role.
Tilt of earth as it circles the sun makes the seasons appear.
Spread of sunlight over earth makes length of days in a year.

SCIENCE IN POETRY

Land and Water Affect Climate
Land and water absorb and release heat at a different rate.
Inland city climate is different from a city by ocean or lake.
The temperature of ocean water changes little from day to day,
But the temperature of oceans affect climate in this way:
If a body of land borders an ocean, the water cools the wind.
So the distance from the equator is not always climate's trend.
Direction of winds in the United States is mostly west to east,
So many cities have climates that are affected by the west.
Water will evaporate from a liquid to a gas.
So the amount of water vapor on our coast is usually vast.
Humidity is what we call this water in the air;
Plants, too, give off water vapor, so water is everywhere.
Bare mountains and hot deserts have little humidity.
The human population prefers some rain, you see.
The ocean currents have a say in temperatures of land.
The cold ocean currents cool the land like a fan.
As icy waters from the poles move inward to a shore,
The ocean winds from arctic lips cool or freeze some more.
Currents from near the equator warm land like a wrap,
England is warmer than Newfoundland though parallel on the map,
That is because Newfoundland has currents from North Pole,
While England has our warm Gulf Stream for a warming role.

Elevation Affects Climate
Climates change with elevation, the heights of land or hill;
Away and up from flatland heat the air will cool or chill.
When vapor is carried by ocean winds over mountain tops
The cooler air condenses and rises in tiny drops.
These tiny drops float on the air to form a snowy cloud.
These drops get bigger as they join, until softly or loud,

SCIENCE IN POETRY

They fall as rain, or snow or sleet; the plan is great
In how the drama of weather knows how to precipitate.
The fog banks move and vapors cool, then drawn up by the sun
Cloud curtains rise, obscuring skies, new cycling has begun.

Reasons for Climate Changes

A volcano can blow a gritty breath
Up through the crust of earth; a quick death
Can come from deadly ash and fumes.
Like Pompeii of old, hot lava consumes
And kills as it flows. The dust in the air
Can hide the sun. The changed atmosphere
Is studied today when volcanos blow.
Constant study by scientists is needed, we know.
We realize that the waste of oil and gas
Must stop if our green planet is to last.
Carbon dioxide trapped in our atmosphere
Scientists warn, can dry the land; so farmlands appear
To be in great danger; polar ice could soon melt.
It gets warmer and dries on our polar belts.
Another outrageous practice, we know, is how
Green forests are victims to saw and plow.
You have shade, you have oxygen, you have beauty
When you carefully plant or attend to a tree.

Learning About the Universe

Everything in space that is served by our sun
Is called our universe; to us our earth is number one.
The people who study space are astronomers:
They are scientists who study space and everything there.

SCIENCE IN POETRY

Stars and other objects give off reflected light
They study them through telescopes mostly at night.
Refracting telescopes find light with lens larger than the eye.
So they can gather far more light for us to study the sky.
Reflecting telescopes gather star age and heat information .
Gathering data is the work of every concerned nation.
Do you recall the mirrors at the carnival
Where you saw yourself short and fat, then skinny as a rail?
Do you think that a future reflecting telescope
Could bring down the energy we need? I do; let's hope.
Colors give clues to heat. When you see red or blue,
You know blue is hotter; gas burners are the clue.
To study space through telescopes is a must, and we
Call the places that house them observatories.

Radio telescopes gather radio waves instead of light.
Gas and many objects in space make waves; static is a blight.
The atmosphere obscures some light, so we have moved above it.
With satellites we probe and snoop and earthlings just love it!
A satellite moves quite fast even with no human being in it.
It can circle our entire globe in only ninety minutes.
Some satellites are sent high,some low,but all have a job to do.
We know they look for forest fires, so they protect us too.
From place to place they scan the face of countries to a dish.
From weather beams to tyrant schemes they serve us as we wish.
I am so glad they keep an eye on us through day and night-
They help keep peace while we sleep, our wondrous satellites!

SCIENCE IN POETRY

A space-probe is an unmanned craft
That studies our universe fore and aft.
As it studies our whole universe,
It sends messages to computers here on earth.
Codes make pictures for us to see;
Spacecrafts make close-ups in our solar sea.

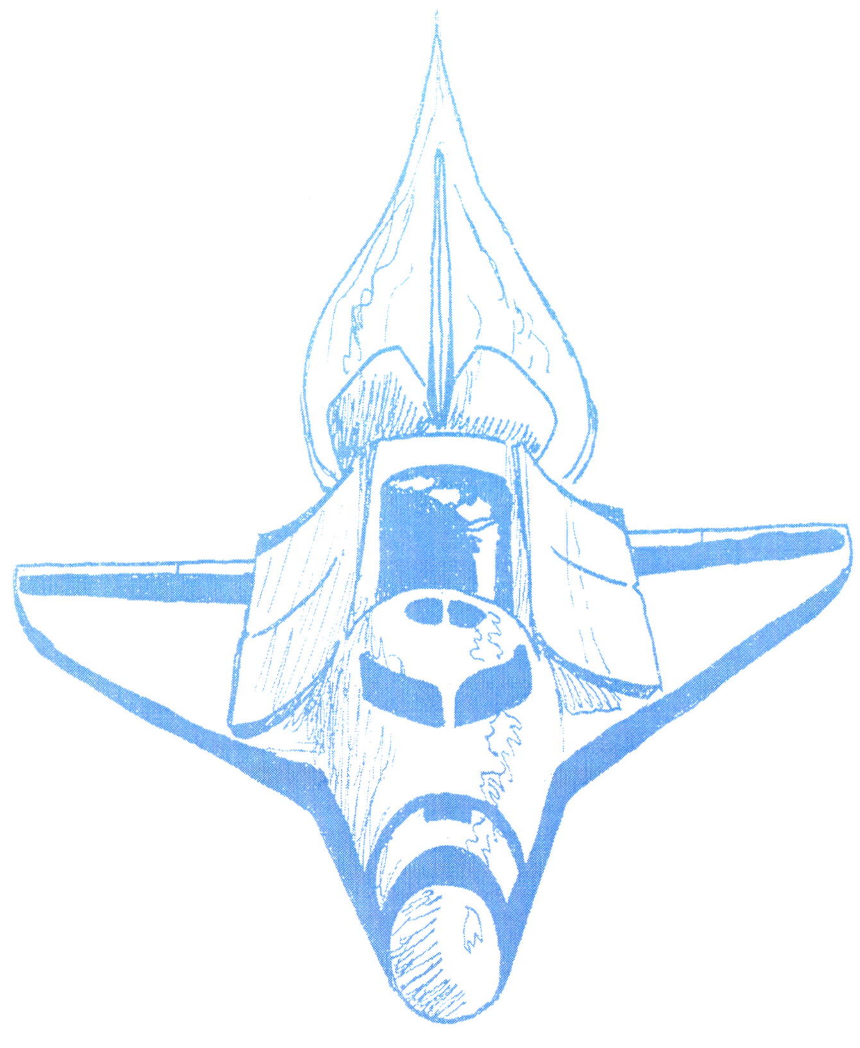

SCIENCE IN POETRY

Children of the Sun

Mercury is number one,
Nearest to the sun.
Venus is number two,
Only daughter, is this true?
Earth is number three,
The only one with a family tree?
Mars is number four,
Romans called him "God of war".
Jupiter is number five,
Largest and the "ruler" of the solar tribe.
Saturn has rings, an "S" for number six.
With his many rings does he play tricks?
Uranus is number seven,
Brightest but unseen in our heaven.
Neptune is number eight, named for the God of the sea.
Pluto is farthest. MVE MJS UNP are our Galaxy.

(Couplets)

Measuring Distance in Space

The average distance between the earth and the sun
Is called an astronomical unit. This unit is only one.
The planet Jupiter is five times farther out.
You can see how measuring planet distance came about.
The distance light travels in one year
Is called a light year. When stars appear
The light you see started at least four years ago.
Light travels 186,000 miles a second, as you know.

SCIENCE IN POETRY

Seven times around the earth in a second, that is fast!
Distances outside our solar system are so very vast.
We know that earth is turning in its orbit run.
Scientists measure distance from different positions.
Astronomers study distance to a nearby star.
Then later in the year they re-read from where they are.

The star has not moved, but Earth has moved:
So with the two readings, distances are proved.
Example may be your thumb held out toward a wall,
Look at thumb with both eyes, it does not move at all.
Now alternate right with left, changed distance takes place,
The reason why: eyes occupy a different place on your face.

Mapping the Stars

Ancient people used the stars to help them navigate.
A map of skies signifies space launching times and dates.
A group of stars together make up a constellation.
Romans, Greeks and Indians gave names to star formations.
Indians thought "Big Dipper" poured blood in the Fall.
The falling and the changing leaves cast a saddened pall.

We have all heard of the zodiac,
Which I associate with "quack".
We will not honor such "stuff" here. . .
Except to say that they are the "signs of the year".
Why one would believe in astrology
Is certainly a mystery to me.

SCIENCE IN POETRY

The sky on summer nights is different from that in winter.
Different star groups come into view, weather will be warmer.
The farmers view stars to know the time to plant a crop.
Wind and rain may bring on change, but seasons never stop.
Stars, dust, and gases make up our Milky Way Galaxy.
These are billions of stars held together by their gravity.
A galaxy is a building block of our universe.
Our milky way is like a maze unknown to prose or verse.
Our solar system is a part of galaxy Milky Way.
How many galaxies exist, no one can really say.

Three things affect the brightness of a star,
Distance, size, and temperature are what these things are.
We call star brightness magnitude, which will depend
On size of star, of distance, and on the light it sends.
The brightness of a star is told by its temperature.
A blue- white star is hotter than a red one, we are sure.
The sun's surface temperature is 5,700 Celsius:
That is 10,292 Fahrenheit! you will check this, I trust.

UNIT IV: THE HUMAN BODY

Mrs Crunde

For you to remember the body systems-
Mrs. Crunde is the dame
She will help you to remember
There are Systems in her name.

****

Muscular System lets us move.
Reproduction System. . . prodigies prove!
Skeletal System is our framing shapes.

Circulatory System: Blood and plasma circulates.
Respiratory System: Our breathing in and breathing out.
Urinary System: Our filters for blood, our waste route.
Nervous System: Our thinking and muscle movement system
Digestive System: Eliminates waste, uses food for nutrients.
Endocrine System: The "Master Glands" of emotional sense

This is an Acrostic poem. Read the first letter and go down the page to spell "Mrs. Crunde."

Bones

Your body can hold its shape because your bones support it.
Without your Skeletal System you could not walk or sit.
Bone cells make hard bone tissue; this tissue is made from
Proteins, the element phosphorus, and calcium.
Tiny tunnels run through bones to carry nutrients.
They also carry oxygen in blood contents.

SCIENCE IN POETRY

The center of bone is hollow, filled with marrow inside.
This soft tissue is a fine diagnostic guide.
Some marrow makes red blood cells, some marrow stores fat.
When you study more anatomy you will see the good in that.

A very hard tissue covers the bone.
Tissues, called ligaments, connects bone to bone.
Cartilage is tissue that protects bone and connects.
In your rib cage it seems elastic- - the way it can flex.
Cartilage is wonderful in substance and in kind:
Without its protection our bones would grate and grind.
The hard bones of your skull protect your organ brain.
They protect eyes and inner ears; the reason is quite plain.
The fragile spinal cord drops from brain through vertebrae.
Brain sends every message through our Nervous System way.

Damage to the spinal cord can cause paralysis.
Muscles cannot move without the nerve assist.
Your rib cage protects large blood vessels, lungs, and heart.
Your pelvic girdle protects; later, you will learn each part.

JOINTS:
Ball and Socket, Hinge, Pivot, Gliding and Fixed

Cup one hand, then make a fist with the other one,
A ball and socket joint example when you make hands join.
Your shoulder and hip joints are both made like these.
Hinge joints move back and forth; your elbow and your knees
Are good examples when you move them back and forth
It is easy for you to see what such movements are worth.
The pivot joint in your neck allows your head to turn, move.
The cartilage between neck bones protects the spinal groove.

SCIENCE IN POETRY

The gliding joints in wrist and ankles are useful indeed.
Think of how you use them to realize the need.
The fixed joints in your skull do not move
After you are born. Suture (^^^^) lines remain and prove
Where cartilage used to be; cartilage could contract.
The center front of baby's head is soft, a fact
That makes it most important to handle baby with care.
Babies are precious so we must stay aware
Of anything that could cause them harm.
We must keep them safe and warm.
Bones of a baby are made up mostly of cartilage.
Bone cells replace this cartilage as minerals engage.

At ends of bones the cells divide so the bone can grow.
Your bones may lengthen to age sixteen, or twenty-one or so. . .
Broken bones will heal as bone cells make new bone tissue.
Even when you stop growing your bones can thicken too.

Muscles

Muscles work with bones and other systems too.
Without Muscular System there's little you could do.
Tendons connect bone to muscle so your muscles can flex.
Bend elbow, muscles in upper arm flex as bottom ones relax.
The muscles in your arms and legs are highly visible.
You watch them bunch or relax: this you can do at will.
These voluntary muscles are the ones you control;
Involuntary muscles play another role.
Involuntary muscles in stomach squeeze to move your food.
Your heart muscle, too, is involuntary, dependable and good.
Voluntary muscles tire more easily;
Involuntary muscle tissue, like the heart, can last a century.

SCIENCE IN POETRY

Keeping Muscles and Bones Healthy

Your entire body has needs and you are very wise
To eat nutritious food, get proper rest, and lots of exercise.
Fish, beans, meat and milk are proteins you need.
Carbohydrates found in grains give stamina and speed.
Vegetables and fruits are important to your diet.
If you neglect any of these foods you are not eating right.
Exercise strengthens muscles, especially the heart.
Sleep is important to your health, a very vital part.
Rest and sleep are healers; your body must relax!
Repairs take place, your blood sheds waste, these are facts.

Cramps, Strains and Sprains of Muscles

A cramp is a very painful. It is a muscle contraction.
If you rub the muscle cramp you may get back into action.
A muscle strain occurs when muscle is stretched out too far.
Usually a light ace bandage is all that you require.
A sprain is far more serious: Tissues are damaged
When twisted around a joint; elevation, an ace bandage,
And rest is then recommended. A sprain of ankle or wrist,
Is caused by a tearing, wrenching twist.

Fractures

A fracture is a break in bone.
Two kinds of fractures are well known.
A simple fracture stays in place;
In compound fractures, bones separate.
A simple fracture may be held in place by a splint.
A compound fracture usually needs surgery. Extent
Of injury dictates a splint or a cast.
In small children, fractures require constraints that last.

SCIENCE IN POETRY

A compound fracture is a serious injury:
Bone breaks tissue, muscles, and blood vessels you see.
Infection is the biggest fear
When compound fractures appear.

How Your Body Grows

There are chemicals in your body made by glands.
These are growth hormones, genetically planned.
The pituitary gland controls how you grow.
Growth spurts differ, as we all know.

Each of us grows at a different rate.
Your birth parents' genes help dictate
How tall you will grow. Nutrition plays a part
Of how healthy you are; give yourself head-start.

Respiration

Emptying and filling your lungs is spoken of as respiration:
Breathing out, expiration; Breathing in, inspiration.
Your respiration is about 12 to 16 times a minute.
You breathe in needed oxygen, as carbon-dioxide exits.

The passage from your nose connects to your throat.
Your voice-box is your larynx, your vocal cords of note.
The larynx sits on top of your wind pipe called the trachea.
The trachea is the tube through which you must breathe.
Your trachea branches into bronchial tubes, before these end,
They further branch to air sacs where gas exchanges begin.

SCIENCE IN POETRY

You know that you could not live
Without the oxygen that you inhale.
Again it's known far and wide
That you would die unless you exhale carbon-dioxide.

Your two lungs are the organs that bring this about.
There is a marvelous exchange when you breathe in and out.
The large muscle that separates chest from stomach cavity
Is called the diaphragm. This muscle helps you breathe.
The cartilage between the ribs help the lungs expand.
It is a fine study to know how you are planned.

How the Body Uses Oxygen
There are 4,000.000 rbc's* in fifteen drops of blood.
Of course this is approximate but generally understood.
Each red blood cell carries oxygen through the arteries
To every cell your body has; You know the need of these.
Color of the blood in arteries is brighter red than veins.
Your veins picked up some waste in cellular exchange.

The water part of your blood is called plasma.
The blood cells float in plasma like tiny ships at sea.
Venous blood carries carbon-dioxide, darker blood as stated,
To heart, through veins, to lungs, to heart, to body, oxygenated.
In life, this constant circulation
Takes place throughout your life's duration.

Your arteries divide into smaller arteries,
Then into smaller vessels called capillaries.
Through these tiny vessels oxygen and nutrients pass.
Your body cells are fed this way; Do they feast or fast?
* Red Blood Cells

SCIENCE IN POETRY

Oxygen Releases Energy

The Digestive System processes food we eat.
The mouth, stomach, and small intestines greet
Each nutrient with processes. Food in blood
Is carried the same as oxygen; Blood sugars are our food.

Broken into carbon-dioxide and water, energy is released.
Cells use oxygen to release energy. Duties are increased;
To grow, divide, fix damage, or just to keep you warm.
The list of needs for energy is longer than your arm.

Getting Rid of Body Waste

The Excretory System gets rid of body waste.
Without this ability we would disappear with haste.
Cell waste is different from waste in the large intestine.
Intestine wastes are solids that come from food ingestion.

The lungs exhale carbon-dioxide, the skin excretes sweat.
Each cell throws off carbon-dioxide, so every need is met.
Cell carbon-dioxide is carried off by plasma in the veins
To lungs, to our inhaled breaths; there's oxygen exchange.

Sweat is water mixed with waste from the sweat gland.
Waste goes out through pores; that's how the skin is planned.
Blood vessels carry salt and waste through tubes to pores;
Pores are tiny openings in the skin which are little doors.
Unless you soap and scrub your skin, bad odors will remain.
Because skin sheds, you would smell like goats after rain.

SCIENCE IN POETRY

Two kidneys filter blood, they are in your lower back.
Ureter, bladder and urethra are your urinary track.
Kidneys are held by arteries and veins sans bone protection.
Kidneys are prone to injury and often blood infection.

Kidneys extract water and salt and blood nitrogen.
Too much nitrogen in your blood will surely do you in.
Kidneys, with their filtering, make urine all the time.
Through ureters to a bladder, this system serves you fine.
This sac has many nerves that tell brain when it's full.
So you can control your water passage as a general rule.

Caring for Your Respiratory and Excretory Systems

You have learned how systems work in close harmony
In getting rid of body waste to keep your body healthy.
In the Respiratory System the nose filters dirt and dust.
When working with air-borne substances, a mask is a must.
Smoke, pollen, and dust may cause allergies;
Allergic reaction is when tissues swell from these.
If your respiratory tissues swell, it's hard to breathe.
Antihistamines, when ordered, help to some degree.

To smoke cigars, cigarettes, or pipes is dumb.
Who is there among you who would want to harm his lungs?
Tobacco smoke causes cancer, a wild growth of cells.
Emphysema is caused by carbon dust and smoking as well.
Emphysema is a condition that harms the lung air-sac walls;
They can't exchange carbon-dioxide for oxygen, this appalls!
Since blood circulates through lungs, cancer cells spread.
So smokers often have lung disease and are prematurely dead.

SCIENCE IN POETRY

Your excretory system may go on the blink
Unless you know that lots of water is a must to drink.
Kidneys remove salts from blood; salt can turn into stones.
Ultrasound can break up stones, as treatment now has shown.
"An ounce of prevention is worth a pound of cure".
This old adage is important; how does this apply here?

Protective padding over kidneys is a sincere need.
If blood shows up in urine, this is serious indeed.
So, respiratory, urinary, digestive, and circulatory
Are all a part of our system we call Excretory.

Teeth

In the study of teeth, the scientists note
That a carnivore's teeth are pointed and sharp.
An herbivore's teeth just would not do
If you have to tear flesh as the carnivores do.
We humans beings have teeth that serve just right
For the animal and plant foods in our diet.
Eight incisors are four front teeth in each jaw.
Their sharp cutting edge can bite meat that is raw.
Our canine teeth are four: two above and two below;
Their purpose is also biting and tearing, we know.

Premolars and bicuspids are eight in our set.
Two behind the canines; they can grind, we can bet.
The teeth in the back are molars: these are twelve.
We must brush our teeth, teeth cannot brush themselves.
Count up the numbers, above and beneath:
You will always remember you have thirty-two teeth.

SCIENCE IN POETRY

****

Kite Flying
dedicated to John Jordan

Yes, I can catch the spirit of the sky
And bring it quivering through a strand of string.
The exultation of it all is why
I find kite-flying such a wondrous thing.
To roll and wheel and dip in fine contempt
Of ocean winds, too fragile to com pare,
I master turns no plane would dare attempt,
Ground pilot to a thing set free on air!
And as my kite begins to clinb the wind,
To ride the string like some poor captured bird,
I want to cut the ties and give it wing,
To send my inclinations heavenward.
   Calm-giver to my hand you often teach
   Something of heart in trembling out-ward reach!

Epilogue

SCIENCE IN POETRY

A Toast to the Scientists Who Are Working on The Twisted Ladder

"A family's heritage may be mundane;
But we have found genetic legacy
momentous in importance to us all.
To trace an illness of calamity
written in our genes from ages past
has, through familial research of like groups,
helped us define keen concepts of ourselves."

The doctor pulled down charts for us to view.
"Twenty-three pairs of chromosomes are here.
This twisted ladder is our legacy.
Some fifteen-hundred illnesses are penned
this day to our hereditary map.
Crisis management and damage control
have been the treatments of our century.
Study and identification lures
us scientists to prevention. Nor can one
ignore the facts of progress we have made.
Our hope lies in relentless study. . .Hours
turn into days and days to years until
a link is found that lends to cell repair.
Finding psycho-pathology brings fear
and consequences of a numbing scope.
To know to wed would bring calamity —
Has it been better to espouse blind chance?

SCIENCE IN POETRY

And some will fear the loosening of reins
on Dogma's throat. Stained windows in the past
have cracked before to let non-judgement in.
A billion cells are spread across our map.
No one escapes his set heredity.

The Maker's pattern will one day be known.
The glass we see through has been dark, indeed.
Wisdom comes from God alone. . . and we
come to discoveries in His own time.
There is no separate trail to Science and Faith.
In me, essence of both lie in the cells.
Dominant genes smash through a thousand years
to cause more wreckage than an errant train.
You have heard of them; Huntington's disease,
Tourettes, Sickle cell anemia and
Cystic fibrosis to name a dreaded few.
We have the markers where some sit at last!
Much to be done to pry them off the rung;
To prevent misery to Earth's family.
Where will the Salem fingers point to place
blame...fifty or a hundred years from now?"
(Blank verse)

SCIENCE IN POETRY

A Bouquet to the Teachers of English

English Harvest

I heard her gentle voice and saw his face:
He did not know just why her lesson plan
Should include him. The subject was old lace,
Trite, unimportant. Just how could the man
He was to be, appreciate the day
When bright and shining lines that he would scan
In English class would straighten and repay
Her dedication to the Poet's plan?
And why was she so zealous to bestow
These singing beauties till they were a part
Of him who heard? Oh, it took time to know.
Boy grows to man and then nuances start

On golden threads of thought that come and go.
So long to wait, so slow to start to grow!
(Sonnet)

INDEX

A

B

Crust, 58
Crustaceans, 26
Cytoplasm, 17

F

Fahrenheit, 45
Ferns, 21
Filaments, 51
Fire, 40
Fish, 28
Fission, 53
Five kingdoms, 15
Fixed joint, 79,80
Force, 52
Fossil, 52
Fossil Fuel, 52
Fractures, 81
Frogs, 29
Fungus, 15,39

G

Galaxy, 77
Gasoline, 52
Gastropod, 25
Gasses, 41
Generator, 54
Genus, 15
Geothermal, 57
Gills, 28
Glaciers, 62
Glands, 82
Gliding joint, 79
Grand Canyon, 61
Granite, 63

Mites, 27
Mixtures, 45
Molars, 86
Mold, 64
Monera, 15
Monocot, 20
Molecules, 44
Mollusks, 25
Mosquito, 28
Mosses, 21,60
Moth, 27
Mrs. Crunde,Body Systems 78
Mud puppies, 29
Muscle cramps, 81
Muscle sprain and strain, 81
Muscular system, 80

N

Nervous system, 79
Neutron, 43
Newton, 52
Nitch, 39
Nitrogen, 43
Noise, 68
Noise pollution, 68
Non-metals, 43
Non-renewable resources, 53
Nuclear energy, 48,54
Nuclear plants, 48,54
Nuclear power, 48
Nucleus, 43

Nutrition, 82

O

Observatory, 73
Ocean current, 71
Octopus, 25
Oil spills, 66
Oxygen, 43,83,84

P

Padding, 86
Paralysis, 79
Parasite, 23
Particles, 45
Pelvic girdle, 79
Periodic table, 43
Perspiration, 16
Petals, 19
Petrify, 64
Petrified Forest, 64
Phosphorus, 78
Photosynthesis, 17,19
Physical property changes, 41
Physical weathering, 58
Pioneer stage, 40
Pituitary gland, 82
Pivot joint, 79
Planarian, 24
Plants, 16,17,18,21
Plates 63

R

Wind, 57,62
Wings, 32
Worms, 23,24

ABOUT THE AUTHOR

Winifred Schramm, author, registered nurse and world
traveler lives in Pasadena, Texas with husband, Arthur. She
and her daughter own and operate a Kwik Kopy Printing
franchise.

She was published in The Knoxville News Sentinel while a
student in Benham High School, Benham, Ky. She co-wrote
a poetry column in The Kentucky Post while a student nurse
at St. Elizabeth Hospital in Covington, Ky. After graduation
she served thirty -two months in Africa and Italy with the
U.S. Army Nurse Corps.   She was commended for bravery
on the Anzio beachhead when the 56th Evacuation Hospital
was bombed.

After the war, she was a private student of F.W. Roe, Ph.D.
(Literature) of the University of Wisconsin at Madison.
Poetry has been a search all of her life. She received a
teaching certificate from the University of Texas in 1965.
She was a private student of Merlin Goldman in his medical
laboratory in Victoria, Texas. Science has been a life-long
love. While on the faculty of Victoria High School, Victoria,
Texas, she created an addition to the school's curriculum to
train students for entry-level hospital jobs. This highly
succesful program, along with her work with student poetry
and drama, earned her recognition as "Teacher of the Year"

in 1979 by the Texas Retired Teachers Association. Janel Hobbs, Home Economics Supervisor, assited with the teaching of this course

A frequent contributor to volumes of poetry published by regional presses, she was most recently published by the University of North Texas. In 1990, she was invited by the Vietnam nurses to Washington, D.C. to represent World War II nurse-poets.She and her husband, Arthur F. Schramm, Lt. Col. Ret., continue their association with students by volunteering as mentors in the H.O.S.T.S* reading program sponsored by the Pasadena Independent School District, Pasadena,Tx.

* Helping One Student To Succeed

Other Works by Winifred Schramm

SONNETS FROM THE CLASSICS, *Can You Name Them?* and Other Poems, published by Wilkins & Associates Inc., Pasadena, Texas., 1995

PINGAZOO :(A Toast to All the Towns That Grew Around A Spanish Square). This word was given as a toast by Brazilian pilots during Word War 11. Published by I.E. Clark, Plays Magic, 1967.

DRAMATIZATION: "The Other Wise Man" by Henry Van Dyke, published by Encore Performance Publishing, Orem, Utah

THE MIKADO GOES WEST, a historical musical for high school student presentations.

Index to Incorporated Poems

Camping in Texas, from <u>Sonnets from the Classics, Can You Name Them? and Other Poems</u>, by Wilkins & Associates, Inc. Pasadena, Texas, 1995.

Mrs. Crunde: Body Systems, used in classroom instruction, Victoria Independent School District, Victoria, Texas

Kite Flying, published in <u>Listen to Texas</u> by Armstrong Publishing, Victoria, Texas

The Twisted Ladder, from <u>Sonnets from the Classics, Can You Name Them? and Other Poems</u>, published by Wilkins & Associates Inc., Pasadena, Texas, 1995

English Harvest, published in <u>Pegasus</u>, by Victoria College, Victoria, Texas

ABOUT THE ARTIST

Taylor Springle is a junior at Friendswood High School, Friendswood, Texas. Three librarians have applauded his cover illustration as a sure-fire eye-catcher for the intermediate student.

His business address is:

> Taylor Springle, Artist
> Baxter Press
> 700 S. Friendswood Dr., Suite B
> Friendswood, Texas 77546
> Phone: 281-992-0628
> Fax: 281-992-0615

His father, Pat Springle, is a well known author and book publisher. Give them a ring. I find them to be superior human beings.

Index of Illustrations